KU-682-776

Looking at Everyday Things

Jean and David Gadsby

A and C Black · London

General editor: R J Unstead
1 Looking at Other Children
2 Looking at Everyday Things
3 Looking at Britain
4 Looking at the World Today
5 Looking at Scotland
6 Looking at Wales

Books 1–4 are also published
in a complete one-volume
edition called *Looking at the World*

The authors are grateful to the following for their help with specific chapters of this book:
Gwynneth Ashby (*A trawler goes to sea*); British Sugar Bureau (*Sugar*); Cadbury-Schweppes Ltd (*Chocolate and cocoa*); Ceylon Tea Centre (*Tea from Sri Lanka*); Danish Agricultural Producers (*Food from Denmark*); Fyffes Group Ltd (*Growing bananas*); K A Halfhide Ltd (*Building a house*); HJ Heinz Co. Ltd (*Baked beans*); W F Henderson (*Wool and wheat from Australia*); National Coal Board (*The coal miner*); National Cotton Council of America (*Cotton*). They would also like to thank Countryside Properties Ltd, K A Halfhide Ltd and Lee Cooper Ltd for their help in obtaining illustrations.

Photographs
Australian News & Information Bureau 63b and c, 66a; Margaret Baker 27a and b, 32b, 33b, 34c, 35a, b and c; Peter Baker Photography 10a and b, 11a; Barnaby's Picture Library 4, 5a and b, 12a and b, 31a, 34a and b, 44a and b, 45a, 54c, 61; BBC Photographs 53b; BP Oil Ltd 75b and c, 76a and b, 77a and b, 78a and b; British Sugar Bureau 13a and b, 14a, b and c, 15a and b; British Sugar Corporation 16a, b, c and d; British Trawlers Federation 50b; Cadbury-Schweppes Ltd 17a and b, 18a and b, 19a, b and c, 20a, b, c and d; Canadian High Commission 6a and b, 30a; J Allan Cash Ltd 29c; Ceylon Tea Centre 71a and b; Christopher Cormack cover, 70a and b; Courtaulds Ltd 28a, b and c; Danish Agricultural Producers 3, 41a and b, 42a, b and c, 43a and b; Jeremy Finlay 9b; John Free 13c; Walter Fussey & Son 46c; Fyffes Group Ltd 21a, b and c, 22a, b and c; HJ Heinz Co. Ltd 56c, 57b, 58a, b and c, 59a and b; W F Henderson 62b, 63a, 64b, 65b, 66b; Alan Hutchison Library 72a; International Wool Secretariat 62a, 64a, 65a; Lipton Export Ltd 72b and c, 73a and b; S C Mason 51a; Massey-Ferguson Industries Ltd 7a and c, 8a and b; Michigan Bean Shippers Association 56a and b, 57a; National Coal Board 36, 37b, 38a and b, 39a and b, 40a and b; National Cotton Council of America 1, 23a and b, 24a, b and c, 25a and b; Overseas Containers Ltd 51b; R K Pilsbury 52a, b, c, d and e, 53a and b; Popperfoto 30b; Practical Gardening 54a, and b.; Ranks Hovis McDougall Ltd 9a; Simrad and Decca Radar Ltd 48b; White Fish Authority 47a and b, 48c; Barnaby's 7b.

Drawings Brian L Ainsworth.
Designed by Ann Samuel.

All rights reserved. No part of this publication may be reproduced, stored in any retrieval system, or transmitted, in any form or by any means, electronic, mechanical, photocopying, recording or otherwise, without the prior permission of A & C Black (Publishers) Ltd

Fourth edition 1980 Reprinted 1981
Copyright © 1980 A & C Black (Publishers) Ltd, 35 Bedford Row, London WC1R 4JH

Printed and bound in Great Britain by Hazell Watson & Viney Ltd, Aylesbury, Bucks

British Library Cataloguing in Publication Data
Gadsby, Jean
    Looking at everyday things.–4th ed.–
    (Looking at geography; book 2).
    1. Geography–Text-books
    I. Title   II. Gadsby, David   III. Series
    910   G128

    ISBN 0–7136–1953–8
    ISBN 0–7136–1851–5 (non-net)

Contents

About this book

It is Saturday morning. The Bell family is going shopping. Mr and Mrs Bell, Peter and Susan get into their car and drive to the big town near their home.

When they reach the town centre, Mr Bell, Peter and Susan take a big trolley round the supermarket, stocking up for the week. They collect bread, biscuits, bacon, eggs, butter, cheese, sugar, tea, fish fingers, baked beans, a frozen chicken and washing powder. Mr Bell also picks up a can of oil for the car.

Meanwhile Mrs Bell is looking for fruit in the market. She buys apples, oranges and bananas.

The Bell family meets again outside the carpark and put the shopping in the boot of the car. Then they all visit a big store to buy new cotton tee shirts and jeans for the children, a woollen cardigan for Mr Bell and some new tights for Mrs Bell. On the way home they stop at a garage to buy petrol. Peter and Susan are hungry, so Mr Bell buys some chocolate too.

Every day the Bell family needs many things from all over the world. This book tells you about some of these things. It tells you how nylon is made and how fish are caught.

It tells you how coal is mined and how sugar and cotton are grown. It tells you of many other everyday things, and about the way in which they come to our factory and shops.

In this book you will also read about hills and valleys, about winds and the weather, about rivers, maps and the shape of the earth.

B5

Wheat from Canada

Wheat grains from which flour is made

A prairie farm at harvest time. The trees protect the farm buildings from strong prairie winds

Most of us eat bread at almost every meal, and we know that we can always go to the shop and buy a fresh loaf. Millions of loaves are eaten in Britain every day. Loaves are made from the seeds of a plant called wheat. Where does wheat come from?

Britain does not grow enough wheat to make all the bread that we need. We have to buy wheat from countries which have plenty of farmland and can grow more wheat than their own people need.

Wheat from Canada

Much of the bread made in Britain is made partly from wheat grown in Canada. In the main wheat-growing area of Canada, the prairies, the winters are very cold, but the summers are hot and fairly dry, and the wheat grows easily. Because of the cold winter most Canadian farmers sow their wheat in spring. It is called "hard spring wheat".

The farms are very large and there are no fields as we know them—just wide stretches of land growing wheat. Machines are used for most of the work so only two or three workers are needed on each farm.

The winter is a quiet time on the farm. The land is covered by a thick blanket of snow for four or five months.

At the end of the winter the snow melts and the ground thaws. The farmer prepares the soil by running a cultivator over it. This cuts the weed roots just below the ground. He does not use a plough or turn the soil over. The prairie winds are so strong that they can blow loose soil away and ploughed land dries out more quickly than unploughed.

In early May the farmer fastens a seed drill to his tractor. The seed drill cuts a furrow and drops seed and fertilizer into it. The rows are 15 to 17 centimetres apart.

In June the farmer may have to spray his land with weed killer.

In the spring the farmer uses a cultivator to prepare the soil

Using the seed drill

Loading the seed drill with seed

B7

Wheat from Canada

Towards the end of August the earliest-sown wheat is ready to cut. The farmer uses a swathing machine which cuts the wheat about 24 centimetres above the ground. The wheat is left lying on the stubble until it is dry enough to keep safely in store for several years. Then, about the middle of September, the farmer begins combine-harvesting.

The combine-harvester is a machine that cuts the wheat and threshes it. Threshing means taking the seeds or grains of wheat out of the ear. In some parts of the prairies the farmers can cut and thresh the wheat at the same time, but most farmers swathe the wheat first.

The combine-harvester empties the wheat grains into a lorry which drives along beside it. The lorry takes the wheat to the nearest railway station and dumps it in a pit. From the pit it is lifted up into a grain elevator, a tower beside the railway line. The wheat is stored in the elevator until it is taken by train to the port to be shipped overseas. Some of the wheat comes to Britain.

The swathing machine cuts the wheat and leaves it lying in rows to dry

The combine-harvester picks up the swathed wheat and threshes it

In a big bakery the loaves move through the oven on a conveyor belt and come out at the other end, crisp and brown

Preparing dough for the oven in a small bakery

In the flour mill steel rollers crush the wheat grains and turn them into flour.

At the bakery the flour is mixed with salt, water and yeast. The yeast is a tiny plant which feeds on flour. As it feeds, it makes new yeast plants. Yeast gives off a gas which is trapped in the dough and makes little holes in it, like a sponge. As more and more holes are made, the lump of dough grows bigger and bigger. We say that the dough is rising. When the dough has risen to twice its size, it is baked in a hot oven.

In a big modern bakery huge machines look after every stage of bread-making. But sometimes old-fashioned bakeries still make bread and you may be able to visit one and see the baker kneading the dough, baking the bread in an oven and bringing out crisp browned loaves.

B9

The story of a river

A stream high in the mountains

It is a fine sight to see a stream in the mountains after a rainstorm. The water rushes along, splashing over the rocks. Sometimes it makes little waterfalls.

The stream gradually makes a narrow, steep-sided valley for itself. It flows so quickly that it carries stones and mud along with it. When the stones are first broken off by the stream, they are sharp and jagged. But the rushing water whirls them on, crushing them against each other and wearing them down until they are round and smooth.

If you find round, smooth stones in your garden, you can be sure that they were once pebbles in a stream.

At the end of winter, if the snow melts quickly, the stream becomes a torrent. In dry weather the stream is just a trickle or it may even dry up.

Water rushing along a rocky stream bed

A river flowing through a wide valley

Some of the rain which falls on the hills does not flow into a stream. It sinks into the ground. If it reaches a layer of solid rock which it cannot pass through, it runs along the top of the underground rock until it finds a way out. Then it bubbles out of the ground as a spring. This is another way that a stream is made.

Above is a river which started as a mountain stream. Other streams, called *tributaries*, have joined the river, making it larger. Now the valley is wide, and the hills on either side are gently curving. We can row boats and swim in this part of the river, for the water flows much more slowly than it did in the mountains.

No rocks stick out of the water and the river is not flowing fast enough to move large stones. But it is still carrying along fine mud and pebbles.

At bends in the river the current wears away the outer bank. Near the inner bank, where the water moves more slowly, the sand and gravel sink to the river bed.

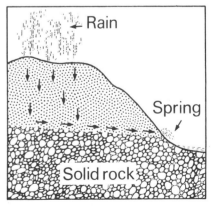

This is how a spring occurs

Flooded water meadows

A dredger. The mud is being emptied into a barge

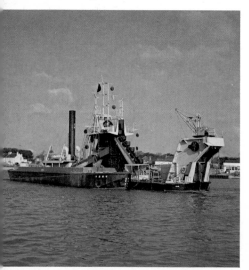

When the river reaches the flat land, it winds slowly to the sea, so slowly that it cannot carry even the finest mud, which it drops on to the river bed.

When there is sudden heavy rain or snow, the river cannot carry away all the water from the hills, and the fields on either side are flooded.

When the flood water has gone, these "water meadows" are covered by fine mud. This makes the grass grow well so that there are heavy crops of hay and rich grazing for cattle.

Some rivers have deep channels when they reach the sea, and ocean-going ships can anchor in them. Quays are built where there is calm water so that ships can unload cargoes.

Sometimes there are mud banks at the mouth of a river. Then *dredgers* have to keep a channel clear. River pilots go on the big ships to steer them clear of the mud banks which are marked by buoys.

B12

Sugar

The sugar we buy in the shops comes from plants. Most plants make some sugar. When the sun is shining, they make sugar from water and from carbon dioxide, a gas which animals breathe out. Plants store the sugar they make as a food and use it to help them to grow. But, in the case of two plants which make a lot of sugar, we take the sugar before the plants can make use of it.

Cutting sugar cane

Sugar cane

Sugar cane is a very tall, strong grass which looks rather like bamboo. The canes have long, spear-shaped leaves and feathery grey flowers. The sap inside the cane is very sweet. Sugar cane grows only in hot, wet places such as parts of India, Africa, Brazil and the West Indies.

How is sugar cane grown? First, a short piece of cane is cut from an old plant. This piece of cane is buried in the ground. New shoots grow up from the joints and soon a spiky leaf shows through the soil. The hot sun and the rain make the sugar cane grow quickly. Sometimes it grows as much as 2.5 centimetres a day.

This machine picks up the sugar cane and loads it on to trucks

Hauling the cane to the factory

Using a machete

Sugar

The cane grows for more than a year. When it is three to five metres high, it is ready for cutting. Men use big, sharp knives called *machetes* to cut through the cane, close to the ground. The green top and leaves are chopped off and used as cattle fodder. The root is left in the ground and will grow again the following year.

When all the cane has been cut, it is loaded into trucks and taken to the sugar factory. There the canes are crushed by huge rollers and sprayed with hot water. The sweet juice which this produces is boiled in big pans until crystals form in the syrup. The mixture is poured into a machine, rather like a huge spin-drier, which spins and separates the sugar crystals (raw sugar) from the syrup (molasses). The raw sugar is loaded into special ships called sugar carriers and sent overseas. The syrup or molasses goes to make cattle food, baker's yeast and rum.

Heaps of crushed cane

Boiling the juice in big vats or evaporators

B14

Beet sugar

About half the sugar we eat comes from countries which grow sugar cane. The rest is produced from sugar beet, a plant rather like a large beetroot with a fat white root. It grows well in cool countries. In Britain sugar beet is grown mainly in East Anglia, Lincolnshire and also in the West Midlands.

The seed is planted in the spring. In the autumn, when the roots are fully grown, a special machine digs up the beet and cuts off the tops which are used as feed for farm animals. The beet are taken by lorry to the sugar-beet factory.

A sugar beet

Sugar-beet factories are open only from mid-September to January but during that time they work day and night without stopping.

The beet are washed and sliced and then mixed with hot water. The sugar from the beet passes into the water to produce a sugar solution which can be treated in much the same way as sugar-cane juice. It is boiled in big pans to form crystals and syrup. The crystals are refined to make sugar and the syrup is used to make industrial alcohol and animal feed.

Harvesting sugar beet

The control panel of the evaporators in which sugar is made from beet

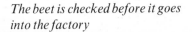

The beet is checked before it goes into the factory

Packing sugar in bags

A tanker of liquid sugar is unloaded at a soft drinks factory

Refining sugar

The raw sugar is brown and, although it has been clarified and filtered at the sugar factories, it still contains impurities and some syrup. At the refinery the syrup is removed and the crystals are dissolved in water. This solution is clarified and the brown colour is removed. A clear, colourless liquid is left.

Some of this liquid is crystallized by boiling. Large crystals are needed for granulated sugar and smaller ones for caster sugar. Icing sugar is made by grinding sugar crystals to make a fine powder. Some crystals are pressed into moulds to make cube sugar. Most of this sugar goes to the shops.

Factories which make soft drinks, food products and medicines may take their sugar as a liquid. It is delivered in tankers.

Sugar is not used for food only. It is also a raw material from which such things as plastics can be made. It can be used instead of oil to make detergents. If a detergent made from oil gets into a river or into the sea, it may kill plant and animal life in the water. But sugar-based detergents do not harm natural life.

Chocolate and cocoa

Chocolate and cocoa are made from the seeds or beans of the *cacao* tree. Many of the beans used by British chocolate and cocoa makers come from Ghana in West Africa.

Kofi has a small cocoa farm. Most of the cacao crop in Ghana is grown on small farms owned and worked by men like Kofi.

The cacao tree likes heat, heavy rainfall and shade, so it grows well in tropical forest. The farmers thin the forest and plant the cacao trees in the shade of taller trees.

Kofi and his son in their cacao plantation

The trees begin producing fruit when they are three to four years old, reaching their peak after ten years. They continue to fruit until they are forty to fifty years old. They grow to about five to eight metres high. The fruit is a pod with a leathery rind just over a centimetre thick. When it is ripe the pod turns yellow.

A cacao pod

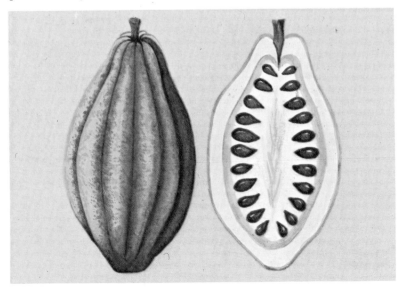

Inside the pod are the cacao beans

B17

Spreading the beans in the sun to dry

Opening a sack of beans at the chocolate factory in Britain

Chocolate and cocoa

The main crop is picked from the end of September to the beginning of January, with a much smaller crop from May to July. At harvest time Kofi cuts the lower pods from the tree with a machete or cutlass. To cut down the higher pods he uses a knife on a long stick.

Kofi splits the pods in half. Inside are cacao beans in a sticky pulp. He takes out the beans and pulp and covers them with banana leaves to protect them from the rain. He leaves them for nearly a week. The pulp "sweats" or ferments and runs away, leaving the beans. During this sweating the chocolate flavour develops.

The beans are then spread on tables to dry in the hot sun for several days. Then they are poured into sacks and taken by lorry to the nearest port. There they are loaded into the holds of cargo boats and exported.

Making chocolate and cocoa powder

When the sacks of cocoa beans reach the cocoa and chocolate factory, they are sieved and cleaned. Then they are roasted in revolving drums. This brings out the aroma, the pleasant chocolate smell, and makes it easier to remove the shell and to break the beans into small pieces or *nibs*.

More than half the cocoa bean is made of a fat called *cocoa butter*. When the nib is ground, the heat of grinding melts the cocoa butter and a sticky brown liquid called *cocoa liquor* is produced. When it is cooled, the liquid sets solid.

To make cocoa powder the manufacturer needs to *remove* some of the cocoa butter from the liquor. To make chocolate for eating he needs to *add* cocoa butter to the liquor.

Cocoa butter is removed by heavy pressure. The hard dry cake that remains can be ground up to make cocoa powder. When this has been packed in tins or packets, it is ready for sale. Some of the cocoa is mixed with sugar and is packaged as drinking chocolate.

There are two kinds of chocolate for eating: milk and plain. Most of the chocolate eaten in Britain is milk chocolate.

B18

Chocolate and cocoa

One way of making milk chocolate is to mix cocoa liquor, milk and sugar together. The milk used is full-cream milk which has been condensed to remove some of the water which it contains.

The chocolate mixture is dried in large ovens and comes out as large pieces of *chocolate crumb*. These pieces of crumb are then ground up and mixed with extra cocoa butter to form a paste.

To make the chocolate smoother the paste is ground between heavy rollers. Then it is beaten in a machine for many hours to produce the best flavour.

The liquid chocolate is still not ready to be made into bars. It has to be *tempered* by mixing and cooling to get its temperature and texture just right. This is the point at which fruit and nuts may be mixed in.

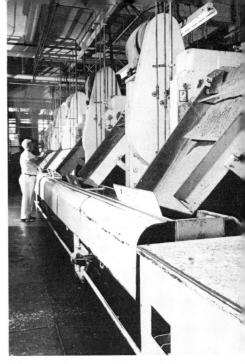

The chocolate crumb is ground between heavy rollers

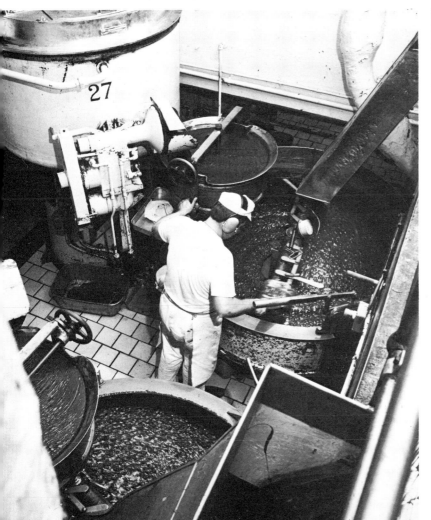

The liquid chocolate is poured into moulds

Mixing nuts and raisins into the chocolate

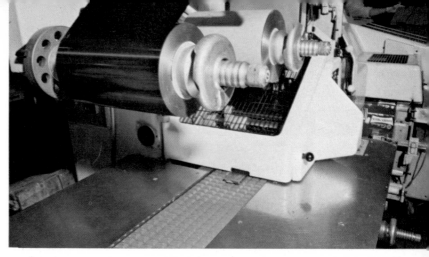

Bars of chocolate entering the wrapping machine

Packing boxes of chocolate into cartons

Chocolate bars are made by pouring liquid chocolate into moulds which are moving along in an endless chain. The moulds are then shaken to spread the chocolate evenly and to get rid of any air bubbles. The moulds pass through a cooler to set the liquid chocolate. Finally the solid bars of chocolate are tipped out of their moulds on to a moving belt which takes them to a wrapping machine.

Individual chocolates are made by first shaping the centres in small moulds and pouring liquid chocolate over them and underneath them. Then they are cooled and packed into boxes.

Plain chocolate does not contain milk. It is made by mixing cocoa liquor, sugar and extra cocoa butter. The mixture is then made into chocolate bars in the same way as milk chocolate.

Decorating chocolates

Packing chocolates into boxes

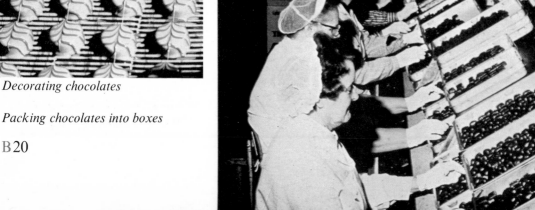

Growing bananas

The bananas in our shops all come from tropical countries in West Africa, Central America and the islands of the Caribbean. Banana plants need the right soil, sunshine and steady rainfall. Farmers make sure the plants have plenty of water by irrigating the fields with canals and ditches and by using spray systems that make an artificial rainstorm.

A bunch of bananas

Banana plants can grow to ten metres high but the farmers prefer varieties that grow to about four to five metres: they don't blow down so easily in tropical storms.

Bananas are grown from pieces of banana root. These "seed bits" send up new shoots and make a new plant. After about nine months the plant produces a single flower which develops into a giant bunch of a hundred to two hundred bananas. While the bananas are growing, the plants are sprayed by aircraft to protect them from disease and insect pests.

The bananas are not picked until a ship is available to take them overseas. Then the farmer is told how many bunches of bananas he should cut.

Hooking a bunch of bananas on to a conveyor belt

An irrigation ditch which brings water to the banana plantation

Washing the bananas

Weighing the bananas

The "backer", the man who does the cutting, uses a machete to cut through the stem just below the big bunch of bananas. He swings the heavy bunch, which weighs from 10 to 30 kilograms, on to his back.

The bunches are taken to packing stations where each one is washed and cut up into smaller bunches that can be packed into cartons, ready for shipment.

When the bananas are packed, they are still hard and green. They must stay like that during the voyage, so the banana ships are specially designed to keep the bananas cool.

When the bananas arrive in Britain, they are unloaded on to insulated railway vans and lorries and taken to ripening centres. They spend four to five days in specially heated rooms until they are ripe enough to be repacked and sent to the shops.

Packing bananas

The bananas you see on sale are usually a clear yellow. If you keep them at home for a day or two, little brown spots appear on the skin. This means that the bananas are fully ripe and ready for eating.

B22

Cotton

A cotton boll and an unripe one which has not yet burst open

Have you got a tee shirt or a pair of jeans? If you look at the label inside, you will probably find that they are made from cotton. Did you know that cotton comes from the seed pod of a plant?

Cotton grows in warm lands which have rain in the spring, followed by a warm dry summer. One of the best-known cotton-growing countries is the United States of America.

The cotton-growing states stretch right across the south of the country from the Pacific Ocean in the west to the Atlantic Ocean in the east.

The cotton farmer plants the seed as soon as he is sure that the winter frosts are over. This can be as early as February 1 in the south of Texas or as late as June in the north of the cotton-growing area.

He uses a mechanical planter which opens up several furrows at once and drops seed into them. It then smoothes the earth over the seeds. At the same time the machine spreads fertilizers to help the plants grow and also spreads chemicals to kill weeds.

When the young plants come through, they must be thinned out and kept free of weeds which would use the water and food that the young plants need. Careful spraying with weedkiller and the use of a machine called a cultivator get rid of the weeds.

Planting cotton

Insect pests, particularly one called the boll weevil, can damage the cotton crop. So the cotton fields are sprayed with insecticide, a chemical that kills insects. This is usually done from an aircraft.

B23

Young cotton plants

A cotton flower

Cotton

About two months after the seed is planted, flowers appear on the cotton plants. When they open, they are cream-coloured, but they change colour from cream to pink and then to dark red. When they fall, a seed pod is left behind. This is called a cotton boll.

Inside the cotton boll are the young seeds. Moist fibres grow from the seeds. As the boll ripens, the fibres grow and grow until they split the boll open. The fibres burst out of the boll in a white fuzzy mass. Chemicals are used to remove the leaves from the plant so that the sun can reach the bolls and ripen them more quickly.

The cotton bolls are harvested by machines. Each machine has sets of spindles. As these revolve, they entangle themselves in the cotton boll and pull it off the plant. One machine can harvest a tonne of cotton in an hour.

Harvesting cotton

The cotton bolls are sent to a factory called a *gin*. The cotton is sucked into the building through pipes. A machine called a *gin stand* tears the fluffy cotton, the *lint*, from the seeds.

The raw cotton is pressed into hard-packed bundles called bales. The bales are wrapped in sacking and held firm by metal bands. Then they are sent to cotton mills in the United States or overseas.

At the cotton mills all the tiny fibres of cotton are cleaned and combed by large machines. Then they are spun or twisted to make one long thread or yarn. This can be woven or knitted into cloth.

The cotton seeds are not wasted. They are still covered with fluff, so first of all these tiny fibres or *linters* are removed. These can be used for making rayon, a man-made fibre used for making yarn and cloth.

Then the seeds are crushed for the oil they contain. This is used to make margarine and cooking oil. It is also possible to make a very nourishing flour from cottonseed. This can be used to increase the food value of other flours and to take the place of expensive meat in such things as tinned stew.

Piles of cotton waiting to be treated in the cotton gin

A big pipe sucks the cotton into the cotton gin

B25

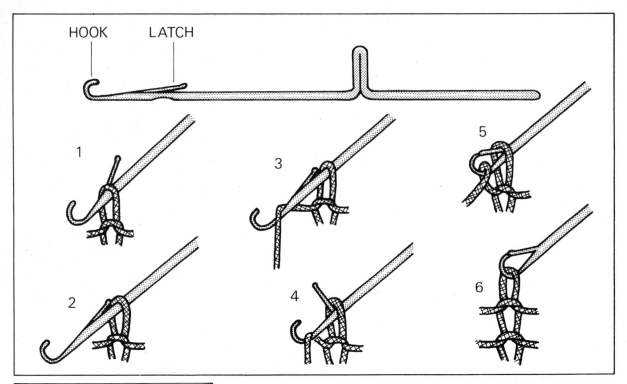

How the needles in a knitting machine work

The cotton for jeans and other clothes made from cotton fabric is woven on looms. The cotton for a tee shirt is knitted.

You have probably tried knitting using knitting needles. But have you used a knitting dolly? The stitches are on pegs and you make new stitches by laying wool against each peg in turn and lifting the old stitch over it.

This is how a knitting machine works. Instead of pegs it has a row of hooked needles, rather like crochet hooks, except that each one has a latch. The machine pushes the needles forward and the stitches on them slip behind the latch. The machine lays yarn across the empty hook. The needles are pulled back again and the old stitch slips forward, closing the latch and sliding off the now closed hook. This can be done very quickly.

A knitting dolly

Cotton

Cotton cloth is woven in many different thicknesses. Some cotton materials are very fine and you can see through them. Others, for overalls and jeans, are very thick and hard-wearing.

You may have seen your mother cutting out a dress. She places the pattern pieces very carefully on the material which has been folded in two. Then she cuts round the pattern.

This is what happens in a clothing factory too. But, instead of using scissors to cut through two layers of material, the cutter in a factory uses a special machine which cuts through many layers of fabric.

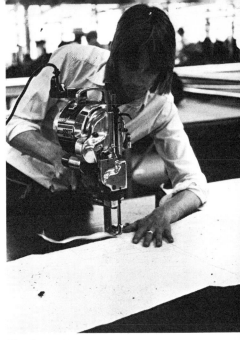

Cutting out cotton jeans

In a factory that makes cotton jeans powerful sewing machines are used to stitch the jeans together. Each machinist does one part of the work, sewing on a pocket, for instance, and then passes the work on to someone else who stitches the leg pieces together or sews in the zip. Then the jeans are pressed and sent to the shops.

Stitching jeans together

Nylon

Nylon chips

How many things made of nylon are you wearing? Underwear, sweater, socks, shirt, blouse?

Nylon does not grow on a plant like cotton or on an animal like wool. It is made from chemicals by men and machines.

The chemicals used are made from oil and coal. At the end of a long process a hot sticky substance is dropped into cold water and hardens into a solid lump. This is cut into small pieces which look like chips of white marble. This is nylon.

To make a thread or yarn that can be knitted or woven, the nylon is melted again. It is forced through tiny holes. As each tiny stream of liquid hardens, it becomes a fine thread. When this is stretched and twisted, it can be used like wool or cotton to make cloth.

Nylon threads being formed

Weaving nylon into cloth

The shape of the earth

Our world is a huge ball spinning in space. For many years it was believed that the earth was flat, and long ago there were sailors who would not sail far from land because they were afraid of falling off the edge of the world.

It is difficult for us to believe that the earth is round, but we can prove it to ourselves if we look through a telescope at a ship coming towards the land. This is what happens:

1   The smoke appears over the horizon
2   We see the funnel and top of the ship
3   The whole of the ship can be seen

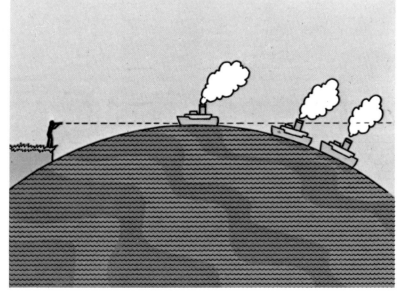

The drawing shows how the ship comes over the horizon. (The earth is not really curved as much as this, of course.)

Astronauts in space can see the earth as a ball. This is how the earth looked to the men on board a spacecraft. Arabia is at the top of the globe. Can you pick out Africa?

Near the North Pole ships can sail on the Arctic Ocean only during the summer

At the Equator, in Kenya, the weather is hot throughout the year

The shape of the earth

Light and warmth from the sun

Have you noticed how much colder it gets on a winter afternoon when the sun sets? This is because the sun is giving less warmth (as well as less light) to the earth.

The higher the sun is in the sky, the more heat reaches us. When the sun is low in the sky, we receive much less heat. At the Equator (an imaginary line round the middle of the earth) the sun is almost overhead throughout the year. So it is hot there every day.

At the North and South Poles the sun is never very high, even in the short summer, so it is very cold. In the winter the sun does not rise above the horizon at all.

About halfway between the Equator and the two Poles the sun is high in the sky in summer and low during the winter. So the summers are hot and the winters cold.

Britain lies in this region. But our weather is not quite like this. Winds from the sea (cool in summer and warm in winter), and the clouds they bring, keep our climate mild.

On the Canadian prairies, the same distance from the Equator as Britain, the winters are bitingly cold and the summers very hot. This is because they are so far from the sea and the winds that reach them have crossed land (icy cold or burning hot), not sea.

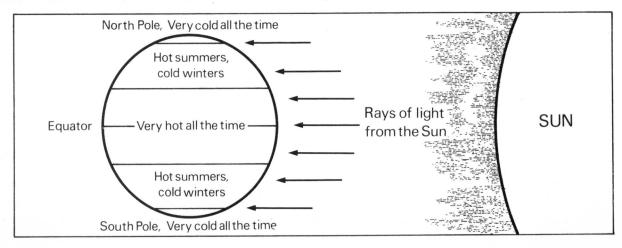

North Pole, Very cold all the time

Hot summers, cold winters

Equator — Very hot all the time —

Hot summers, cold winters

South Pole, Very cold all the time

Rays of light from the Sun

SUN

Building a house

When a new estate of houses is to be built, the first job is for a surveyor or architect to make a careful plan of the *site*, the plot of land where the houses are to stand. He shows the exact size and shape of the site and where the nearest drains and electricity, gas and water supplies are.

The architect looks at the site

An architect designs the houses, making drawings for the builder to use. He makes a drawing of each floor in a house, showing the size of the rooms and where all the doors, windows, fireplaces and fittings are. This kind of drawing is called a *plan*. He also makes drawings which show what the house will look like from the front, the back and the sides. This kind of drawing is called an *elevation*.

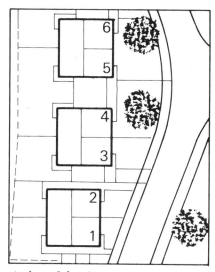

When the architect has completed his drawings, they are given to the quantity surveyor. He measures the drawings and then writes down how many cubic metres of concrete, thousands of bricks, metres of timber and other materials are needed to build the house. He also lists the numbers of doors, window frames, drain pipes, baths, radiators, electric sockets and other supplies which will be fitted into the buildings.

A plan of the site

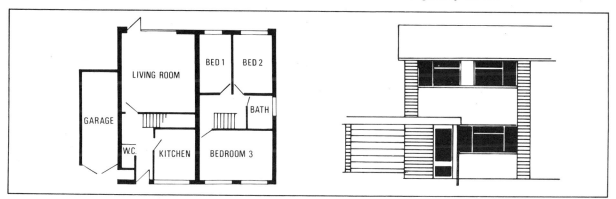

Plan (it shows the ground and first floors)

Elevation

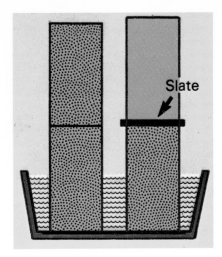
How a damp course works

The bricklayer at work on the outer layer of the wall

Building a house

The foundations

To make firm foundations for the houses, trenches are dug out with excavators and filled with ready-mixed concrete. The bricklayers begin to lay bricks for the walls on this base.

Just above ground level the bricklayer lays a damp course. This is a layer of waterproof felt. Water cannot pass through it and so damp cannot rise up the wall.

To see how the damp course works, stand two bricks in water, one on top of the other. Water will slowly creep up both bricks. But if you put a slate between the bricks, water cannot pass through it, and the top brick will remain dry.

Can you find the damp course in your house? Look at the outside wall. Is there an extra thick layer of mortar between the bricks about 20 centimetres above the ground? That shows where the damp course is.

The foundation work is finished by spreading hard core (rubble or stones) between the walls, with concrete on top.

The walls

Now the bricklayers set to work to build the walls for the ground floor of the house. The wall is made in two layers, an outside one of bricks and an inside one of bricks or blue-grey insulating blocks. The cavity or space between the two layers is 10 centimetres wide. It stops damp from working its way through the wall. Sometimes the cavity is filled with glass wool to keep heat inside the house.

How a brick wall is made

B32

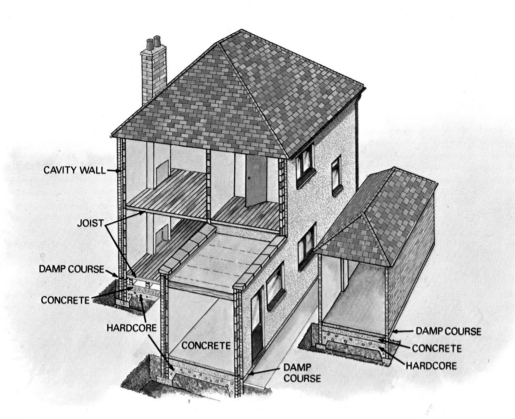

CAVITY WALL

JOIST

DAMP COURSE

CONCRETE

HARDCORE

CONCRETE

DAMP COURSE

DAMP COURSE

CONCRETE

HARDCORE

As the walls grow higher, door and window frames are built in. When the walls reach first-floor level, joists—thick lengths of timber—are laid across the house from wall to wall. These will support the floor of the upper storey.

The bricklayers continue the walls until they reach roof level. Now the roof trusses are hoisted into place. These are giant triangles of wood. The longest side of the triangle is laid across the top of the walls. The two shorter sides of the triangle are the rafters which will form the roof.

When the trusses have been fixed into place, the rafters are covered with waterproof paper to keep out rain and drifting snow. Battens, narrow strips of wood, are nailed through the paper to the rafters. The tiles for the roof are laid on these battens. There are little holes at the top of the tiles so that they can be nailed to the batten.

Inside the house the internal walls are being built. These are made of insulating blocks and are about 8 centimetres thick.

The roof trusses in position

B33

The electric cables will be covered by plaster.

The final stage of plastering

Laying flooring on the joists

First fixing

Now the roof is on, the carpenter, plumber and heating engineer can do the first part of their work.

The electrician puts in his cables, running them through grooves cut in the walls and placing them so that they will be hidden by floors and skirting boards. The heating engineer and the plumber lay their pipes. Gas, electricity and water supplies are laid on.

Glass is placed in the windows by the glazier. Plasterboard is fastened to the underside of the joists to make ceilings.

The carpenter fits window sills, hangs the doors and lays flooring on the joists in the upstairs rooms. The staircase arrives ready-made and is fitted into place.

Plastering

The plasterers lay a thin layer of plaster all over the brick and block walls and over the plasterboard of the ceiling. Now the inside of the house begins to look like a house.

Second fixing

The plumber and the heating engineer come back. The plumber fits the sink unit, wash basin, lavatory, bath and hot-water cylinder. The heating engineer installs the boiler and the radiators. The electrician wires up all the switches and sockets, puts up lighting fittings and fits an immersion heater in the hot-water cylinder.

The walls of the bathroom and kitchen are tiled. The ground floor is covered with plastic tiles. The carpenter fixes cupboards in the kitchen and finishes all the woodwork jobs, fitting fastenings and locks to doors and windows.

Finally the painters come. They paint all the woodwork inside and outside the house. Walls are covered with wallpaper or emulsion paint. A layer of glass wool, 13 centimetres thick, is laid over the joists in the loft to keep heat from escaping through the roof.

A final clean and the house is ready for its new owners.

Fitting radiators

Fixing a switch for the electric light

Tiling the bathroom

The coal miner

Coal comes from under the ground. Many millions of years ago there were thick swampy forests over our land. As the trees died, they formed a spongy mass. Mud and sand slowly covered this spongy mass and hardened into rock. As the rock pressed down hard on the old forest layer, the buried trees were changed into the hard black stuff we call coal. This took about three hundred million years.

Here is a coal mine. Under each of the big towers there is a hole called a *shaft*. Mines always have two shafts in case one gets blocked in an accident. Air for the miners underground is sucked by a large fan down one shaft, through the mine and up out of the other.

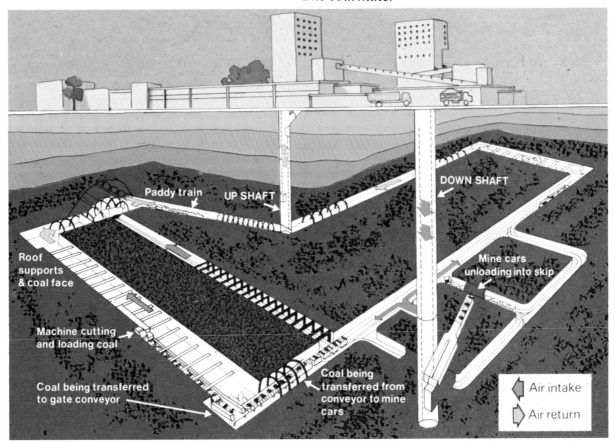

Paddy train   UP SHAFT   DOWN SHAFT

Roof
supports
& coal face

Mine cars
unloading into skip

Machine cutting
and loading coal

Coal being transferred
to gate conveyor

Coal being
transferred from
conveyor to mine
cars

Air intake

Air return

This is a diagram of a coal mine. The picture shows the
buildings at the top of the pit, the two shafts and the seam of
coal.

Work in the mine goes on for twenty-four hours a day. There
are three shifts, each of eight hours. Ted Bates is one of the
miners on the morning shift this week. As he starts work,
the men of the night shift are going home. When Ted
finishes work, the men of the afternoon shift will take over.

When he arrives at the pit Ted changes into his working
clothes. Why do you think he wears a helmet and boots with
steel toe-caps?

Ted always leaves his matches in his locker with his street
clothes. This is because there are sometimes dangerous gases
underground in the mine. These might explode if anyone lit
a match.

Miners in their working clothes

B37

The miner's cap lamp is connected to a battery on his belt

Miners on the paddy train

The coal miner

At the lamp cabin Ted collects his cap lamp. He clips this to the front of his helmet. It runs off a battery pack which he fastens to his belt. Between shifts the batteries are recharged in the lamp cabin.

Before he enters the cage, he hands in a tally, a metal disc with his number on. If an accident happens down the mine, the manager can look at the tallies and see at once how many men are still underground.

The cage is quickly lowered on its strong steel rope and in a few seconds Ted is hundreds of metres below the ground, far away from the sunlight.

The place where Ted works is a long distance from the bottom of the shaft. First he takes a trip on the *paddy train*, a little train just big enough to take twenty men. The train takes him part of the way but he may have to walk the rest: one to two kilometres.

At the end of his walk Ted reaches the *coalface* where he works. This is a wall of rock, usually about 185 metres long, with the coal layer or *seam* in it. The coal seam is about 1.5 metres thick. At each end of the coalface is a tunnel or *roadway* which leads back to the shaft.

Ted is in charge of a big cutting machine which moves along the coalface, slicing the coal away from the seam. At the same time it loads the coal on to a conveyor which carries the coal to one end of the coalface. Here the coal is transferred to another conveyor which runs along the roadway. At the end of this conveyor the coal is loaded into small railway wagons, called *mine cars*, which are then hauled by diesel locomotive to the bottom of the shaft.

The roof of the coalface where Ted is working is supported by many powerful roof supports. These are telescopic and act rather like a car jack. As the coal-cutting machine passes each roof support, a miner moves a control lever. The support automatically lowers a little, moves forward and rises up again to support the roof of the area just cleared of coal.

As the roof supports move forward, one after another, the roof they were supporting is allowed to collapse.

The roadways which run from the ends of the coalface back to the shaft do not collapse. As the miners slowly move across the coal seam, taking the coal and leaving the rock behind, they make the roadways longer. They carefully support the roof with steel arches and walls of stone.

A coal-cutting machine

The miner adjusts the telescopic roof supports. He is wearing knee pads

B 39

The coal miner

An AA van being filled up with petrol made from coal

When the coal has been lifted up the shaft, conveyor belts take it to the *washery*. Here it is cleaned and sorted into sizes by machine. Then it is loaded into railway wagons and taken to the places where it is to be used.

About three-quarters of the coal from Britain's mines goes to power stations which make electricity and to the iron and steel industry. At steelworks the coal is baked in huge ovens to make *coke*. Coke burns at a much higher temperature than coal and is used to make iron and steel.

Many of the jobs that coal used to do—driving trains and ships and making coal gas for cooking—are now done by oil and natural gas.

Coal is heavy and difficult to handle and it is dirty. But the world's supplies of oil may be running out by the end of the century, and natural gas may be finished by about AD 2030. Luckily Britain has at least three hundred years' supply of coal.

Changing coal extracts into liquids that can be used to make chemicals

Scientists are now looking for new cleaner ways of using coal and of making things from it. Work is already being done on making oil, gas and chemicals from coal. It seems likely that in the future coal will be used to make petrol for our cars, gas for cooking and many of the things, such as plastics, nylon and detergent, that are now made from oil.

Food from Denmark

Every week we eat butter, bacon and cheese. These foods may come from farms in Britain but they may also come from Common Market countries such as Denmark.

There is plenty of good farming land in Denmark. It is mostly divided into small family farms which raise pigs, dairy cows and beef cattle. The fields are used to grow grass, barley, oats and root crops to provide feed for the animals.

The cows graze in the fields during the summer. In winter they live in the cow house. There are two main breeds, the reddish-brown Red Danish and black-and-white Jutlands. One cow can give more than 5,000 kilograms of milk each year.

Red Danish cows being fed

A Danish farm house

Food from Denmark

Many farmers milk their cows three times a day. At milking time the cows are driven into their stalls or "standings". They are milked by electric milking machines. The milk is cooled and taken to the village dairy.

The dairy is owned jointly by all the farmers in the village. It is spotlessly clean and well equipped with modern machinery.

To make butter the cream is separated from the milk and then turned round and round in a large stainless-steel churn. Special butter-making machines are also used; these can make between two and three tonnes of butter an hour.

The butter is packed in foil and most of it is sent to Britain. It is sold in our shops as Lurpak or Danelea butter.

To make cheese the milk is poured into a vat like a giant bath tub—it holds about 5,000 litres. Rennet is added to the milk to curdle it. This turns it into crumbly but solid curd and waterlike whey. The whey is drawn off and the curd is pressed into moulds lined with cheesecloth. The cheeses are carefully stored until their full flavour develops.

There are many varieties of Danish cheese—the Danes even eat cheese for breakfast. Danish Blue and Samsoe are two which you will see in your local supermarket.

Churning butter

Cheese-making: (right) the curds after the whey has been drained off; (left) testing a cheese for flavour

To the farmers their pigs are even more important than butter and cheese. The farmers feed their pigs with skimmed milk and whey from the dairy. When the pigs are heavy enough, they are killed for bacon.

Refrigerated containers of Danish bacon and butter being shipped to Britain from Esbjerg

The bacon is "cured" by being put into a bath of brine or salt water. Although brine is wet, the salt in it dries the meat out and preserves it.

Danish farmers breed pigs which look exactly alike—"like peas in a pod"—so that the sides of bacon are all the same size and shape

After brining the meat is stored at a low temperature while it matures and turns into bacon. When it is ready for export, the sides of bacon are loaded into refrigerated containers and taken by road to the port of Esbjerg for shipment to Britain. Nearly half the bacon eaten in Britain comes from Denmark.

Danish farmers produce far more bacon, butter and cheese than the Danes need for themselves. Some of this food goes to other Common Market countries—Germany, France and Italy—but most of it goes to Britain. In exchange Britain sells coal, iron, cars, tractors and machinery to Denmark.

B43

Winds and directions

Trees near the sea often have strange shapes because on most days the wind blows in from the sea. A strong wind makes huge waves which crash over the sea wall. Sometimes the wind blows the waves so fiercely that houses are damaged and the land is flooded.

A weather vane points in the direction the wind is coming from

The direction from which the wind is blowing helps us to know what sort of weather to expect. In Britain, in the winter, winds from the south and west are warmer than those from the north and east. These winds have blown over the sea which keeps warmer than the land in winter. Winds from the north and east have been blowing across land covered with ice and snow.

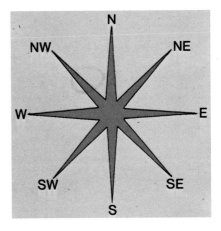

Look for the weather vanes on the tops of high buildings. The arms of the weather vane point north, south, east and west. An arrow above them swings round and points to where the wind is coming *from*.

When we describe a wind we need to know more about its direction than just north, south, east and west. The direction halfway between south and east is described as *south-east*. The diagram shows eight points of the compass. What is the direction halfway between north and west?

Draw a wind rose to show which way the wind is blowing. Each day look at the weather vane and fill in one box to show where the wind is coming *from*.

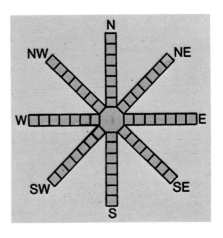

This wind rose shows that the wind has blown for two days from the west, for three days from the south-west and for two days from the south.

B45

A trawler goes to sea

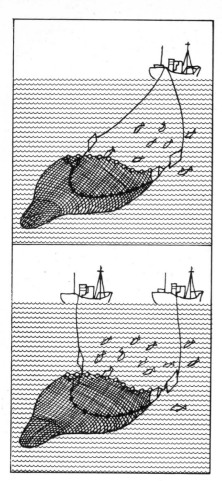

Ocean Star is a multi-purpose fishing boat sailing from a port on the East Coast of Britain. She is called a multi-purpose boat because she can carry out several kinds of fishing.

Most of the time she is *bottom trawling*. Her net is dragged along the sea bed for the fish that swim on the bottom—cod, haddock, coley, plaice and whiting. Sometimes she changes to a *mid-water trawl* net to catch sprats, herring and mackerel. These fish swim between the bottom and the surface. If she wants to use a bigger net, she goes *pair-trawling* with a sister ship. The two ships tow the net between them.

A multi-purpose fishing boat

Ocean Star fishes only in the North Sea. Sometimes she goes a short distance from her home port. Then her five-man crew (skipper, mate and three deckhands) may be away from home for just a night and a day. At other times she goes far out into the North Sea. Then her crew may be away for a week or more.

They cook and eat their meals in the galley behind the wheelhouse. There is a metal framework round the cooking stove so that, when the ship rolls, the saucepans cannot slide off. The table is screwed to the floor and cannot move.

Below the galley is the cabin, with bunks built into the bulkhead or wall. The bunks have sides so that the men sleeping in them are not thrown on to the floor in rough weather.

When the ship is out fishing, the crew may spend as much as eighteen hours a day on deck. They work in the cold wind and salt spray, catching fish for us to eat.

The end of the net, the cod end, is hauled over the side of the boat

Boxes of fish are lowered into the fish hold

B47

- 🟫 Store
- 🟦 Ice
- ⬜ Fish Hold
- 🟦 Cabin
- 🟥 Fuel Bunkers
- ⬜ Engine Room
- 🟦 Bridge
- ⬜ Galley
- 🟫 Crew's Quarters

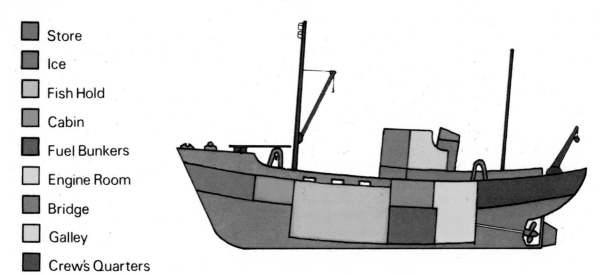

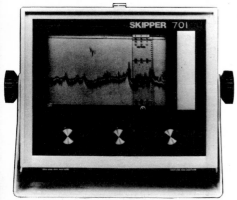

The echo-sounder screen. The wavy lines show the sea bed. The fish are the dark blobs near the top

The skipper uses his VHF radio to talk to the shore or to other boats

Although the *Ocean Star* is a small ship (18 metres long and 5 metres wide), she has a powerful engine. She has a second engine, called an *auxiliary*, which provides power for all her equipment. The wheelhouse is full of electrical and electronic equipment. To help the skipper navigate there is radar, and a radio so that he can talk to the shore and to other ships.

When the *Ocean Star* reaches the fishing grounds the skipper watches the sonar and the echo sounder. The sonar shows where the fish are swimming ahead of the *Ocean Star*. Then, while the boat is trawling, the echo sounder helps the skipper to be certain that he is passing over a shoal. The fish below the boat show as dark patches on the screen of the echo sounder.

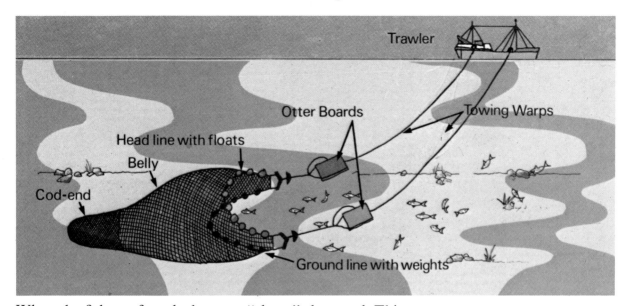

When the fish are found, the crew "shoot" the trawl. This is a cone-shaped net which is pulled along behind the ship. The bottom of the net rolls along on "bobbins"; the top is held up by floats; and the mouth is kept open by wooden or steel doors. These doors are called *otter boards*.

For three hours the *Ocean Star* pulls the trawl through mid-water or along the bed of the sea. The fish swim from the main part of the net through a small opening into the *cod end*. Once the fish are in the cod end they cannot escape.

When the skipper thinks that the trawl is full, he orders it to be pulled in. The winch slowly winds in the wire rope. The mate pulls a knot at the cod end and the fish fall on to the deck—sole, plaice, coley, haddock, halibut and whiting. The mate reties the cod end, the trawl is put over the side and trawling begins again.

Fish caught by bottom trawling

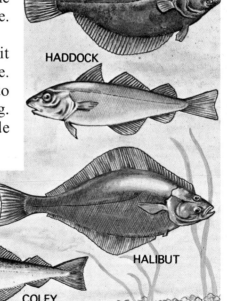

B 49

A trawler goes to sea

Before the *Ocean Star* goes to sea, crushed ice is poured into her fish room. When the fish have been caught, they are gutted so that they will not go bad. Then they are washed and packed in ice in wooden or plastic boxes.

Back at port, the boxes are loaded on to lorries and the fish is sent to all parts of the country. Soon it is on sale in the fishmongers' shops and in the fried-fish shops. Sometimes the fish is bought by the big frozen-food factories. Here some of it will be made into fish fingers and fish steaks.

The *Ocean Star* takes on ice, oil and stores of food and, within thirty-six hours, she is ready for the next trip.

Fish being landed at the port

"Here is the weather forecast"

Several times a day news about the weather is broadcast on radio and television. These are general forecasts which tell us about the weather to expect during the next few hours in our own areas and in the country as a whole. The shipping forecast tells ships at sea about wind speeds, fogs and gale warnings. The farming forecast helps farmers to plan their work and enables fruit growers and market gardeners to protect their crops from frost.

The radio and television forecasts come from the Meteorological Office which collects information about the weather from nearly three hundred weather stations in the British Isles. Special ships in the North Atlantic also serve as weather stations. All kinds of ships act as "weather ships", sending in reports four times a day on weather conditions.

Weather stations send balloons carrying instruments high into the sky. The instruments send back radio messages about the weather at heights of 20 000 to 30 000 metres. Radar stations can track the balloons' movements and learn about the speed and the direction of the wind at great heights.

Radar is used to track weather balloons

This container ship sends weather reports to the Meteorological Office

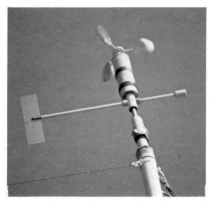

An anenometer measures wind speed

This recorder shows how long the sun shines each day

Storm radar detects coming storms

A barometer

Weather centres all over the world exchange information. Special space satellites send back photographs of cloud patterns and enable weather forecasters to study weather thousands of kilometres away.

The weather stations and ships have special instruments to tell them all about the air—how heavy it is, how much water it is carrying, how hot or cold it is and how quickly it is moving.

The *barometer* tells what the air pressure is, that is how heavy the air is as it presses against the surface of the earth.

To measure how hot or cold it is, that is to find the temperature, the weather stations use a *thermometer*.

On warm days the liquid in the glass tube *expands*, or grows bigger, and rises up the tube. On cold days it *contracts*, or grows smaller, and moves down the tube.

Temperature is measured in degrees Celsius (often called Centigrade) written like this. 10°C. Water boils at 100°C and freezes at 0°C.

In the picture you can see four thermometers. The upright one on the left gives the air temperature. The one on the right shows how dry or damp the air is. The other two thermometers show the highest and lowest temperatures reached during the past few hours.

Cumulus clouds

Cirrus clouds

Weather observers also look at the clouds. They note what kind of clouds these are, how high they are, and which way they are moving.

Here are two kinds of clouds which they often see. The fluffy "cotton-wool" clouds are called *cumulus*. The *cirrus* clouds, which look like a white horse's tail, are often called "mares' tails". They are very high in the sky where it is so cold that they are made of tiny crystals of ice, not water.

Clouds help to tell us about the weather, but they do not give a complete picture of what the weather is doing. So the men and women at the weather stations send reports to the Meteorological Office. There all the information is fitted together.

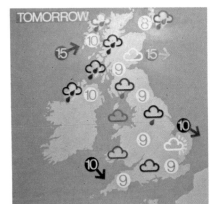

Wind, clouds, temperature, pressure—all these and many other details are put on a map of Europe and the Atlantic Ocean. As the reports come in from weather stations, the map is changed. The forecasters watch the map and take note of the changes. From this they can work out what is likely to happen within the next few hours.

The information for the BBC weather chart comes from the Meteorological Office

Food from the garden

Soon after Christmas Mr and Mrs Bell begin to study the seed catalogues and to decide what crops they are going to grow in the coming year. All the Bells enjoy vegetables, salads and fruit fresh from the garden. As they have a freezer, extra crops are grown for freezing.

Sometimes young plants are bought from a market or garden centre, but usually Mr Bell raises his plants from seed. Some seed can be sown straight into the ground: peas broad beans, carrots, parsnips, cabbage and lettuce, for instance.

Other seeds are sown in pots in the cold frame (a framework covered with glass that protects young plants from cold winds). Tomato plants, runner beans, cucumbers, marrows, leeks and early cabbage and lettuce are usually grown in this way.

Sowing seed

Plants in the cold frame

Planting out cabbage plants

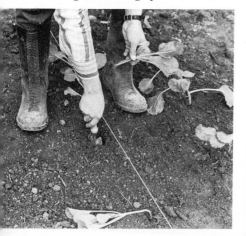

A few days before the tender young plants are ready to be planted out in the garden, Mr Bell leaves the lid of the frame open so that they "harden off" and become used to the wind and weather of the outside world.

The Bells try to make sure that they always have vegetables ready to eat. For instance, by sowing seed every six weeks or so from February to July, it is possible to have cabbages for eating all through the year.

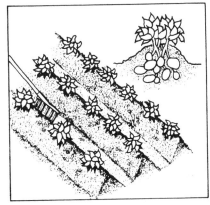

Earthing up potatoes

There is always work to be done in a garden. The soil has to be prepared before it is planted. Mr Bell digs it in autumn so that the winter frosts break up the clods of earth and leave a fine *tilth* of crumbly soil.

He feeds the soil with fertilizer to replace the goodness last year's crops took out of it. He digs in compost to make the soil open and crumbly. Vegetable trimmings from the kitchen, grass cuttings, dead leaves and garden rubbish all go on the compost heap.

During the growing season Mr and Mrs Bell weed the rows of vegetables, water them during dry spells and watch out for signs of disease or insect attack. Some crops need thinning out—carrots and parsnips. Tomatoes need the little side shoots removed. Potatoes must be "earthed up" (have earth piled up round them to keep the lower part of the plant in the dark). Peas and beans need to be trained to climb up sticks or netting.

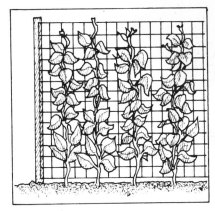

Beans growing up netting

Then in early summer it is time to begin picking. From June onwards the pea pods grow fat. Mrs Bell keeps picking peas, cooking some for the family and preparing the rest for the freezer. She picks strawberries, raspberries and currants and freezes these too or makes jams and jellies.

Other crops are picked in the autumn and stored for the winter. There is not room in the freezer for everything so onions are hung up in the garage. Carrots are packed in a box of sand and apples are wrapped in newspaper and stored in trays. In this way the Bells can eat garden produce all through the year.

Storing vegetables

Baked beans

A Navy bean plant in flower

(Left) The farmer checks the young bean pods. (Right) Hoeing between the rows to get rid of weeds

The Americans invented baked beans but, here in Britain, we eat them much more than the Americans do. We eat 38 000 tonnes a year, about 5 kilograms of baked beans for each person in Britain.

Growing the beans

The beans used for baked beans are Navy or Pea beans, a hard, white bean which grows on a plant rather like the French beans we grow in the garden. Most of the beans come from the state of Michigan in the United States. The flat fertile land is ideal for growing beans. Frosts break down the soil during the severe winter and kill insect pests. The summers are hot for ripening and drying the beans.

The farmer sows the beans in May. By early August the bean plants are 30 to 45 centimetres tall with a white flower. The bean pods grow quickly. The plant turns yellow and the beans are ready for harvesting.

Baked beans

Mechanical pickers pull the plants out of the ground and lay them in rows to dry. Then a combine-harvester gathers up the plants and separates the beans from the plants. The beans are poured through a funnel into lorries which take them to the nearest bean elevator company.

Here the beans are sorted into grades and carefully stored in big silos until they are sold. Then they are packed into bags and loaded, 18 tonnes at a time, into containers (a container is a big wooden box which can be carried on a lorry or railway truck and which fits into the holds of special container ships). The containers are taken to the nearest port and are sent to Britain.

Growing the tomatoes

The tomatoes for making the sauce which gives baked beans their colour and flavour are grown in Portugal or in countries round the Mediterranean. They are specially chosen varieties of tomato with a good sharp flavour and bright colour. As they grow, they are carefully watched and checked.

When the tomatoes are ripe, they are picked and taken to factories. There they are sorted, washed and pulped to make tomato purée. Big five-kilogram cans of purée are sent to Britain.

The combine-harvester picks up the dried bean plants and threshes them

Tomatoes for making purée

Baked beans being inspected as they come out of the oven

Mixing tomato sauce in a giant container

Cooking the beans

The first European settlers in North America learned how to cook the local beans from the Indians—in a pot buried in a hole full of hot stones. The settlers adapted this method. They soaked the beans overnight in a pot, added molasses (see page 14), onions and spices and then put the pot in the oven for the rest of the day. By suppertime the beans were ready to eat.

Sailors found the dried beans a useful food to take to sea but soaking and cooking the beans took a long time. When canning was invented, an American firm began to supply its ships with canned baked beans which needed only to be heated. The long cooking had been carried out ashore.

The baked beans we buy are cooked in a factory. When the beans are unloaded from the containers at the factory, they are cleaned and sorted even more thoroughly, washed and then cleaned again. Then they are soaked in hot water to soften and partly cook the beans before they are baked.

Cans moving along the filling lines

Making the tomato sauce

The tomato purée is tipped into big stainless-steel containers which hold more than 1300 litres and blended with sugar, starch and spices. The spices come from Zanzibar off the coast of Africa, from China, Burma, South America and Madagascar.

Checking the beans for quality and flavour

Filling the cans

The cans and their lids are made in the factory from thin sheet steel covered with tin which does not rust. At the filling lines the cans are filled with the partly cooked beans and then topped up with sauce. Then the lids are sealed on.

Canned food will not keep unless it is sterilized. This means that it has to be treated with heat so that all the bacteria that make food go bad are killed. The cans are sterilized in huge steam-heated retorts, like giant pressure-cookers.

They pass through the retort on conveyor belts at a temperature of about 120°C. Afterwards the cans are cooled, labelled and packed in cartons. Then they are ready to be sent to shops and supermarkets all over the country.

B59

Valley and hill

Here is a valley, with a river running running through it. The old part of the village has a few houses, a church, a school and a little railway station. There is also an estate of new houses, a food warehouse and a factory.

Think about the picture and see how many of these questions you can answer.

1 Where does the water in the river come from?

B60

2 Why is the village in the valley and not on top of a hill?

3 Why is the village near the stream?

4 Beyond the railway is a farm. What crops does the farmer grow, and what animals does he keep?

5 David and Jenny are leaving school and want to work in the village. How many jobs can you find for them?

Here is the moorland, high above the valley. Only sheep can live on the short grass, and they have to wander a long way to find enough to eat.

The grass is short, and hedges will not grow, because there is only a little poor soil over hard rock. Instead of hedges there are stone walls, made of stones which were quarried from the hillsides.

In places where the rain cannot soak through the rock, the ground is wet and boggy. In other places, the rain runs off the hillsides to form little streams.

There are few roads over these hills, only paths and sheep tracks. In winter the snow is often so deep that no one can cross the hills.

Moorland in Wales

Wool and wheat from Australia

A farmer counting sheep with the help of his dogs

Mr Collins is a wheat and sheep farmer in Australia. His farm of 1200 hectares is in flat dry land, covered with mallee scrub. In most years he plants 40 hectares of wheat and uses the rest of his property or farm for grazing sheep.

Although the property is only 300 kilometres from the sea, it usually receives no more than 25 centimetres of rain a year. Sometimes it does not rain for six months at a time. In bad droughts there may be no rain for two years.

Because the area is so dry, water is brought to the farms by channels. This water is not used for growing crops. It is for the animals to drink and it is stored on the farms. The water is only directed into the channels once a year. It is allowed to flow for several weeks so that each farm can store water for the year.

Sheep can live quite well on the dry sun-bleached grass and produce good wool, provided they have plenty of water.

There are two houses on the property, one for Mr Collins and the other for his married son. These two men, father and son, run the whole property by themselves. Only at harvesting and shearing times is some extra help employed.

The channel which brings water to the farm for the animals

Wool and wheat from Australia

Here Mr Collins is "marking" the lambs. He is cutting off their tails, giving them an injection to prevent a disease called pulpy kidney, marking their ears and branding their rumps. There is a tin of special paint at his feet; this is used to brand the fleece.

Once the winter is over and the weather getting a little warmer in August or September (spring months in Australia), shearing begins. Teams of shearers move from farm to farm. They are paid a sum of money for every hundred sheep they shear, so the faster they work the more money they earn. Some shearers are so fast that they can shear more than two hundred sheep in a day. They use electric shears rather like the clippers used by a barber.

Marking the lambs

First they cut the dirty wool from under the sheep's body. Then the rest of the wool, called the fleece, is taken off, all in one piece. Each fleece is thrown down on to the classing table. Here it is *skirted* by men who pull all the dirty and tangled wool from the edge of the fleece. When it has been skirted, each fleece is sorted according to the quality of the wool. Each grade or class of wool is placed in a separate bin until there is enough to press into a bale.

Electric shears

Skirting the fleeces

Checking the quality of the wool at the auction centre

Spray dipping the sheep

Wool and wheat from Australia

The bales of wool are taken from the farms to the auction centre in Melbourne. Buyers come from all over the world to bid for the wools which they need for textile mills. Once most of the Australian wool went to Britain and the Yorkshire mills. Today Japan, the Soviet Union, China and many other countries buy most of the wool clip.

After each sheep has been shorn, it is *dipped*. There are two ways in which a sheep may be dipped. With *swim dipping* the sheep must swim along a narrow concrete trough full of a brown disinfectant. Even their heads are ducked underneath.

Mr Collins uses a *spray dip*. The sheep stand in a tank and they are sprayed with strong jets of disinfectant from the sides and underneath and from showers overhead. It is like being in a shower bath with jets of spray coming from all directions. Dipping is to kill the tiny insects which can burrow into the skin of the sheep.

In drought years there is sometimes not enough grass to feed the sheep on the property. When this happens, Mr Collins sells many of his sheep. But he tries to keep enough to breed new flocks, once the drought has broken.

Using a car to drive sheep

To feed the sheep he has kept, he has the sheep taken droving "on the road". Australian roadways are usually very wide, with a great deal of grass at each side of the hard surface. The girl in this photograph has a flock of almost two thousand sheep which she has been looking after for almost two years. She has a caravan, a truck, a horse and cart and two dogs.

She begins each day by driving the truck and caravan ten kilometres along the road, with the horse trotting alongside. Then she rides the horse back to the sheep and harnesses it to the cart. The rest of the day is spent slowly moving along with the sheep as they graze along the roadside. By late afternoon she has reached the caravan again. The sheep are penned for the night and the horse is fed and allowed to graze.

B65

Combine-harvesting the wheat crop

A wheat silo

The property is also a wheat farm. To sow the wheat seed Mr Collins attaches a combine-seed drill to his big tractor. Superphosphate fertilizer is mixed in with the seed.

The land was prepared some months ago and has been lying fallow (unsown) to keep the moisture in the soil for the growing wheat. Growing wheat when there are only 25 centimetres of rain a year needs a good deal of skill and care. Can you remember what the Canadian farmer on page 7 did to prevent the soil from drying out?

Australian wheat is planted in autumn and grows through the winter and spring. It is harvested in early summer. All wheat is harvested by giant combine-harvesters and is handled in bulk trucks. As soon as it is harvested, it is taken immediately to the silos along the railway track. Long wheat trains take the wheat from the silos to the ports from which it is shipped to many different countries.

B66

Looking at maps

This is a map of part of Crofton village. Peter Todd lives at White Post Farm. When he goes to school he walks down the farm lane and, when he reaches the road, he turns left.

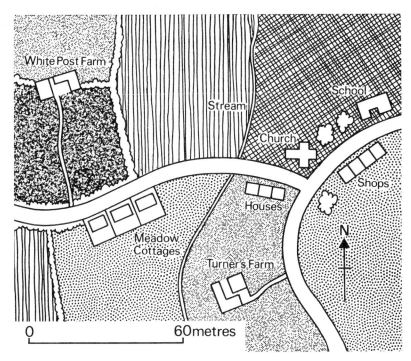

He calls for his friend Robert at Meadow Cottages. They cross the bridge over the stream, turn left at the church and go into the school playground.

After school Peter and Robert walk home. What do they see on their way?

Here is Peter on his bicycle. Will he turn right or left to go to the shops?

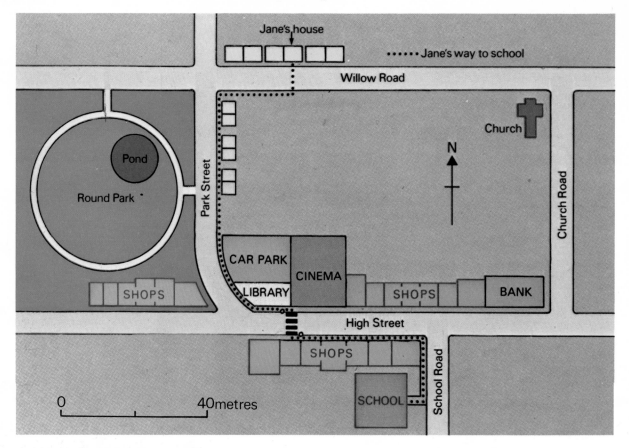

Jane lives in a town. Here is a map of Jane's walk to school. Try to describe the way she goes. Begin: "Jane crosses the road in front of her house and turns right."

Notice that she crosses the High Street at the zebra crossing. Describe another way Jane could go to school.

Point to this building on the map? Is it east or west of the bank?

B68

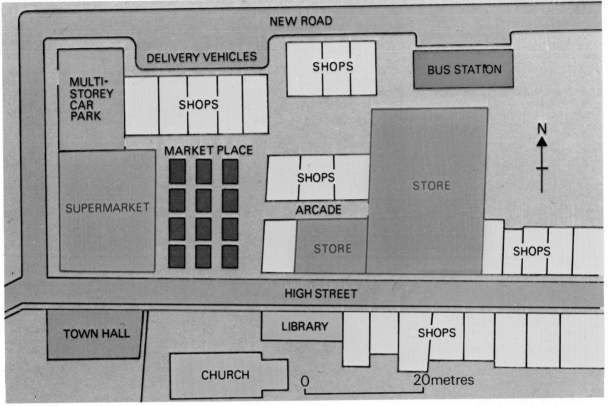

Here is a map of the town centre where the Bell family goes shopping (see pages 4 and 5). You can see the High Street, the church, the library and the town hall, and also the new shopping precinct. You can also see the new road round the outside of the precinct. Lorries can deliver goods straight to the shops.

What other vehicles use the new road and why? Where do they go?

Where does Mrs Bell go to buy fruit and vegetables? Where do Mr Bell, Peter and Susan do their shopping? Where do they meet Mrs Bell?

The last thing they do in the shopping centre is to buy some new clothes. Where do they buy them?

Which is the quickest way for them to take to the shop where they buy the clothes?

B69

Tea from Sri Lanka

The ladies of Queen Anne's reign used to drink "tay" or tea. It cost two pounds a pound, yet they were never sure of the best way to make it. Sometimes they boiled the tea, threw away the liquid and ate the leaves. At other times they made their tea and kept it in barrels like beer.

In Queen Anne's days all our tea came from China. Today tea is also grown in India, Sri Lanka (which used to be called Ceylon) and more than twenty-five other countries. Nearly everyone in Britain drinks several cups of tea every day. If we shared out all the tea used in Britain in a year, every man, woman and child would have thirty packets of tea.

Here is a tea plantation in Sri Lanka. On the hillsides the flat-topped tea bushes are growing. Women in brightly coloured saris are picking the leaves from the tea bushes. Nearby is the factory where the leaves are taken.

Tea bushes grow best on hillsides

The factory building

Planting out the young tea plants

To make new tea plants, cuttings are taken from a big bush. The cuttings are planted in plastic bags filled with soil. After being watered every day for a year, the plants are ready for planting out on the hillsides.

Tea plants like rain but they do not grow well if water collects round their roots. So they grow best on the hillsides where the water drains away.

Tea plants grow all the year round; they are evergreens. If they were allowed to grow fully they would be nearly ten metres tall. But they are pruned so that the bushes are kept down to about one metre in height, with flat tops.

While the bushes are growing, the soil between the rows is hoed regularly to kill the weeds.

The families which work on the tea estates live in small houses, built close together. Because the sun is so hot, many of the houses in Sri Lanka have stone walls with very few windows. To provide shade they have overhanging roofs.

Cuttings

B71

Picking tea

Feeding the tea into the drying machine that turns the leaves black

Sorting the tea into sizes

Tea from Sri Lanka

Many of the workers on the tea estates are Tamils, originally from southern India. They came to Sri Lanka because they could earn more money than in their villages in India.

When the tea bushes are about four years old, the fresh shoots are ready for plucking. In Sri Lanka, which is near the Equator, plucking goes on all the year round. The pluckers are women. Their hands dart quickly over the bushes, nipping off the young leaves and buds.

They throw the tea into the baskets on their backs. As each basket is filled, the plucker takes it to the roadside to be weighed. A skilful plucker can harvest 25 kilograms of green leaf in one day.

When the tea has been weighed, it is taken to the factory. Inside the factory the tea leaves are spread on racks made of nylon. Here they are left to dry for a day. Then the leaves are put in a rolling mill to release the juices in them. A drying machine turns them into the small black leaves which we know. The leaves are sorted, according to size, by sieves.

The sizes have such names as Orange Pekoe, Fannings and Dust.

The tea is sold by auction. All the merchants bid for the tea, saying how much they will pay for it. The one who bids the most buys the tea.

Tea blenders tasting tea

But before the merchant buys the tea, it has to be tasted. The taster makes a pot of tea from each kind of leaf. After six minutes he pours out a cup of tea from each pot. Then he tastes each one in turn by taking a mouthful of tea and then spitting it out again. In this way he judges each tea and decides how much it is worth. He also decides how it can best be mixed or blended with other teas.

Packing the tea in chests

The unblended tea is packed in plywood boxes, called tea chests. Each chest is lined with aluminium paper to keep out the damp and holds about 50 kilograms of tea.

Ships take the chests of tea to countries all over the world. When the chests arrive at the docks, they are unloaded and stored in warehouses. Some of the chests are taken out of store each day. The tea is blended in big revolving drums and is put into cardboard packets and tea bags which we see in the shops.

B73

Oil from the North Sea

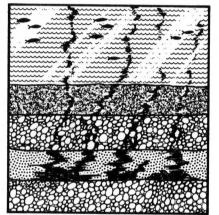

Tiny sea creatures die and fall to the sea bed. They are covered by mud which slowly hardens into rock

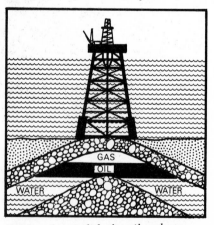

Their bodies turn to oil and gas. Where the rock is spongy, the oil and gas escape to the surface

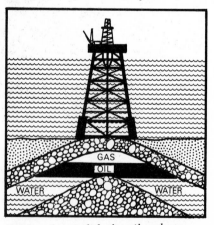

If the rock is solid, the oil and gas are trapped. This is where oil men drill their wells

Until a few years ago all the oil used in Britain came from countries such as Saudi Arabia, Iran and Nigeria. Now much of the oil Britain needs comes from the oil fields in the North Sea.

How was oil formed?

Oil was probably formed millions of years ago, in areas then covered by sea. As tiny sea creatures died, their bodies sank to the sea bed and were covered by mud. Gradually the mud hardened into rock, pressing down on the decaying sea creatures. Their remains were changed to a fizzy liquid made up of oil, water and natural gas—*crude oil.*

Much of this crude oil rose to the surface, passing easily through spongy rock, and was lost. Some, however, was trapped beneath rock it could not pass through. If someone drills a hole through the rock, the fizzy oil gushes out. It is this trapped oil that oil-company geologists look for.

Geologists are scientists who study the earth and rocks. One way to find out about rocks deep beneath the earth's surface is to bore a hole and set off a charge of explosive. The geologists listen to the bang through special instruments and measure how long it takes before the shock waves are bounced back by rocks deep below ground. This tells them what rock formations there are and if they are likely to contain oil.

They do this at sea too but, instead of explosive which could kill the fish, they use an "air gun" which fires compressed air or gas and does not harm the fish.

Drilling for oil

When an area of rock looks as though it might contain oil, a discovery well is drilled. An exploration drilling rig is brought in and firmly anchored to the sea bed. The drilling rig has a derrick, a tall steel tower about 50 metres high. From the top of the derrick hangs the drill pipe. This is made up of 9-metre lengths of pipe joined together. At the lower end is the bit which bites into the rock when the pipe revolves. The bit is made of very hard steel. Some bits have teeth made from diamonds.

When the drill bit becomes worn, the drill pipe has to be hauled up in sections of three or four lengths at a time. That is why an oil rig needs a tall derrick.

If oil is found, the exploration rig is taken away and an oil-production platform is built. From this many production wells can be drilled, thousands of metres deep, so that oil can be taken from a wide area. The wells spread out like the roots of a tree beneath the platform. As soon as one well has been completed, it begins to produce oil, while the drilling crew begin work on another well.

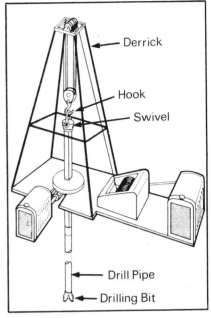

An oil derrick

Some of the drill bits used to bite into the rock of the sea bed

An exploration rig with a supply ship

Drillers at work

The dining room on a production platform

The production platforms in an oil field are linked together by a pipeline which carries the oil to the shore. On land another pipeline takes the oil to a refinery.

Crude oil is a mixture of different grades of oil. Some is thick like tar but other types are very thin. In the refinery all these grades of oil are separated and prepared for their different uses:

> petrol and diesel fuel for lorry and car engines
> fuel oil for heating buildings
> paraffin
> thick bitumen for covering roads.

Some oil products go to chemical works where they are used to make chemicals and materials from which plastics (including nylon chips—see page 28), washing powders, paint stripper and many other things are made.

B76

Living on an oil platform

An oil-production platform is like a small town. As the workers fly in by helicopter, they can see the platform rising out of the sea with its tall derrick. They also see its flare stack for burning off gas in an emergency; its cranes; its three decks of living and working quarters, linked by stairways and a lift; and its helipad.

Nearly two hundred men live there. The drilling crews, of thirty-five men each, work long twelve-hour shifts, changing at seven in the morning and seven in the evening. The men spend fourteen days on the platform without a break and then have fourteen days' leave.

The large quantities of supplies which are needed are flown to and from the platform by helicopter. Extra heavy objects are brought in by sea and hoisted aboard the platform by powerful cranes.

A helicopter taking off from the platform's helipad

A production platform in the North Sea. The drilling rig on the right provides extra space. The ship is bringing in heavy supplies

Part of the metal chamber in which divers live. It is kept at sea-bed pressure

Working on the sea bed

One special group of workers in the oil industry are the deep-sea divers who build undersea pipelines and platforms. At the great depths at which they work, strange things happen to their bodies because of the great pressure of the water.

Normally a diver comes up from a great depth in stages, making long stops at different levels, so that his body grows used to the change in pressure. Without this he risks great pain and even death.

Nowadays North Sea divers are treated rather like astronauts. They spend up to three weeks inside a metal chamber which is kept at sea-bed pressure. They eat and sleep inside the chamber and leave it only to go to work. A diving bell, also at sea-bed pressure, takes them down to the sea bed and brings them up again.

Before they can come out of the chamber, they spend four days resting while the pressure is slowly lowered until it is the same as the outside world.

Divers cannot breathe ordinary air at great depths. They breathe a mixture of oxygen and helium instead. This affects their bodies too. For instance, their voices change and they sound like tape recordings played at too fast a speed!

Locking a diving bell on to the metal chamber. The diving bell takes the divers down to the sea bed

Do you remember?

When the Bells reach home after their shopping trip, Mr Bell and the children bring all the things in from the car and put them in the kitchen for Mrs Bell to put away.

Make a list of all the different foods you can see in the picture (there are at least fifteen). Some of them come from our country, others come from countries in various parts of the world. Against each food write down the names of the countries it comes from, like this:

Butter comes from Britain, Ireland, Denmark, New Zealand...

Look at the labels on other foods and find out what countries they have come from.

Look at the labels on your clothes and shoes and find out what they are made of and where they were made. Make a list like this:

My pullover is made of wool and nylon and it was made in Scotland.

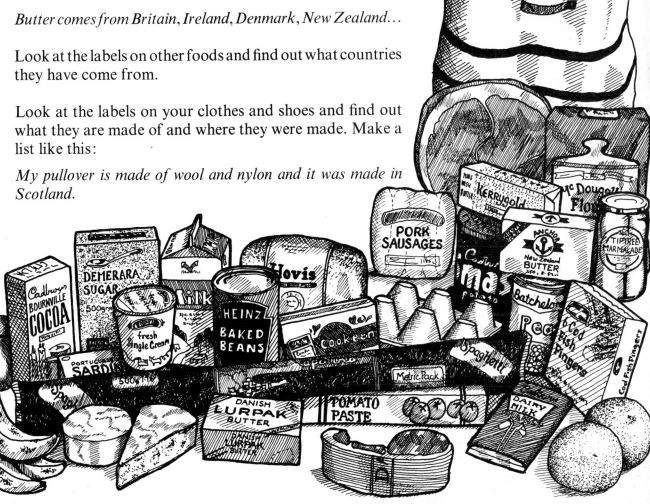

Index

This index will help you look up things quickly.
The numbers refer to the pages of the book.